Animals Have Classes Too!

Amphibians

Jodie Mangor

Before, During, and After Reading Activities

Before Reading: Building Background Knowledge and Academic Vocabulary

"Before Reading" strategies activate prior knowledge and set a purpose for reading. Before reading a book, it is important to tap into what your child or students already know about the topic. This will help them develop their vocabulary and increase their reading comprehension.

Questions and activities to build background knowledge:
1. *Look at the cover of the book. What will this book be about?*
2. *What do you already know about the topic?*
3. *Let's study the Table of Contents. What will you learn about in the book's chapters?*
4. *What would you like to learn about this topic? Do you think you might learn about it from this book? Why or why not?*

Building Academic Vocabulary

Building academic vocabulary is critical to understanding subject content.
Assist your child or students to gain meaning of the following vocabulary words.
Content Area Vocabulary
Read the list. What do these words mean?

* *burrow*
* *carnivores*
* *conditions*
* *larvae*
* *metamorphosis*
* *tentacles*
* *toxin*
* *vertebrates*

During Reading: Writing Component

"During Reading" strategies help to make connections, monitor understanding, generate questions, and stay focused.
1. *While reading, write in your reading journal any questions you have or anything you do not understand.*
2. *After completing each chapter, write a summary of the chapter in your reading journal.*
3. *While reading, make connections with the text and write them in your reading journal.*
   a) *Text to Self – What does this remind me of in my life? What were my feelings when I read this?*
   b) *Text to Text – What does this remind me of in another book I've read? How is this different from other books I've read?*
   c) *Text to World – What does this remind me of in the real world? Have I heard about this before? (News, current events, school, etc....)*

After Reading: Comprehension and Extension Activity

"After Reading" strategies provide an opportunity to summarize, question, reflect, discuss, and respond to text. After reading the book, work on the following questions with your child or students to check their level of reading comprehension and content mastery.
1. What are some of the things that amphibians have in common? *(Summarize)*
2. Why do we know less about caecilians than other types of amphibians? *(Infer)*
3. Where do amphibians live? *(Asking Questions)*
4. What amphibians live near you? *(Text to Self Connection)*

Extension Activity
Pick three to five amphibians you are curious about. Find out more about them. Pretend that you are a scientist. Divide the amphibians into two or more groups based on the ways they are alike and different from each other. Explain your classification system to a friend.

Table of Contents

American bullfrog

Let's Classify!

Look at this animal. What is it? How do you know?

Frogs come in many shapes and sizes. But they all share certain traits, such as long back legs and big eyes.

Scientists group animals and other living things by what they have in common. This is called classification. It helps us identify and understand all sorts of living things.

Classifying Living Things

There are seven main levels of classification. From large to small, they are: Kingdom, Phylum, Class, Order, Family, Genus, and Species.

Animals can be grouped at different levels. Groups based on having one or two things in common are very big. These big groups can be divided into smaller and smaller groups. The smallest groups have only one kind of animal in them.

Animal Groups

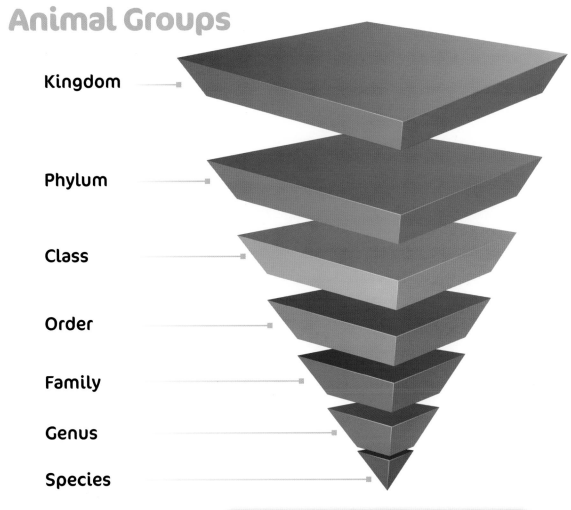

Kingdom

Phylum

Class

Order

Family

Genus

Species

Taxonomy

The science of sorting, describing and classifying living things is called taxonomy. It was first developed by a scientist named Carl Linnaeus.

Carl Linnaeus (1707–1778)

Scientists group all living things into kingdoms.
Animals make up one kingdom.

Frogs are part of the animal kingdom.

Each kingdom is divided into smaller groups. Each of these smaller groups is called a phylum. Frogs and other **vertebrates** are in the same phylum. It is divided into classes. The classes are fish, reptiles, birds, mammals, and amphibians.

Amphibians are different from animals in the other classes. They have no scales, feathers, fur, or hair.

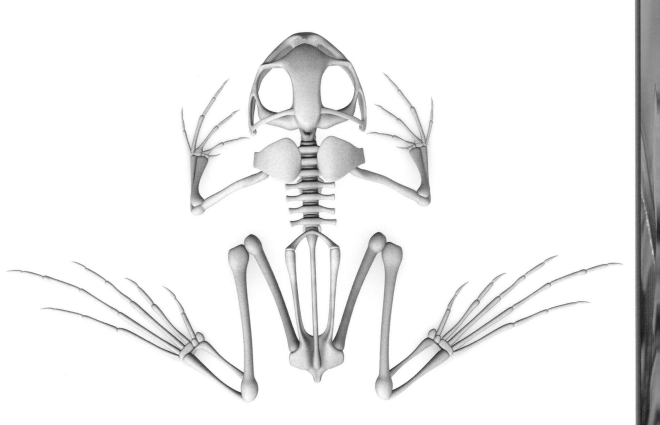

Animals with backbones are called vertebrates. They are in the phylum Chordata.

There are more than 7,800 types of amphibians. All amphibians have certain things in common. All go through **metamorphosis**. This is a series of changes to their bodies. It turns them into adults.

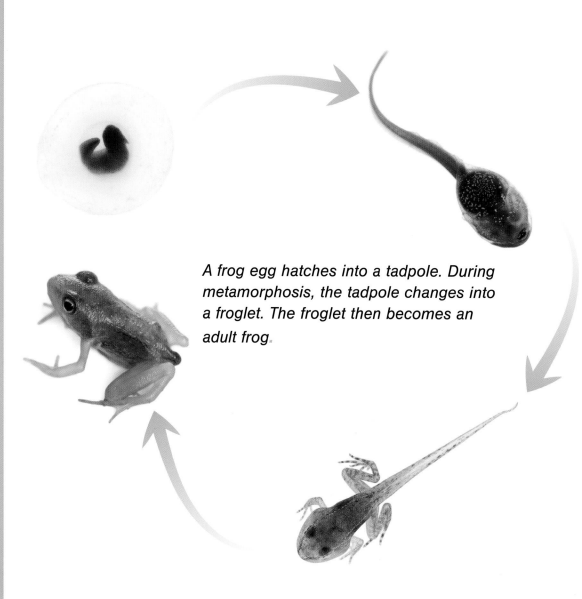

A frog egg hatches into a tadpole. During metamorphosis, the tadpole changes into a froglet. The froglet then becomes an adult frog.

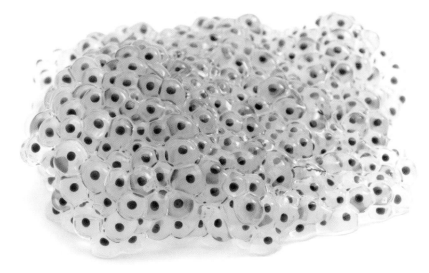

Many frogs lay their eggs in bunches called spawn.

All amphibians are cold-blooded. This means their body temperature depends on their surroundings. Amphibians lay eggs without shells. And all adults are **carnivores**.

Take a Breath

Most amphibians have smooth, slippery skin. Many amphibians can breathe through their skin. Many also have lungs.

Most amphibians live part of their lives in the water and part on land.

Amphibians are found in almost every part of the world. Most live in or near fresh water. This includes lakes, streams, ponds, swamps, and wet caves. Some live in forests, mountains, or even deserts.

Most amphibians lay their eggs in water to keep them from drying out.

If it gets too hot or too cold, amphibians become inactive. They stay this way until **conditions** get better.

One place amphibians can't live is in salty ocean water.

Some frogs can survive freezing and thawing.

Quick to Grow

Spadefoot toads lay their eggs after it rains. Tadpoles become adults in just seven days, before the rainwater dries up.

Frogs and Toads

There are three main groups of amphibians. These groups are called orders. Frogs and toads make up one of these orders. Frogs and toads have short bodies, big heads, and big eyes. Their back legs are strong. Their back feet are webbed.

American toad

Most frogs sing to attract a mate. Males and females gather in ponds in spring. Females lay clusters of eggs. Frog **larvae** are called tadpoles. They live in the water and eat plants. Adults often live on land. They may eat insects, spiders, snails, mice, or even small birds.

Meal Time
Most frogs have long sticky tongues to catch their prey. They swallow their prey whole.

A toad is a type of frog. Toads are different from other frogs. Most frogs have upper teeth. Toads are toothless. Their back legs are shorter. They are less active. Their skin is dry and they live on land.

In the spring, toads travel to ponds to breed.

Toads tend to lay eggs in long chains instead of clusters. Some give birth to live young.

paratid gland

The cane toad has poison glands behind its ears.

Toxic Skin

Many toads have glands in their skin that make a **toxin**. If another animal eats some of the toxin, it can become sick or die.

Salamanders and Newts

Salamanders and newts make up another order of amphibians. All have long tails. Most have four short legs. Most adults live on land. But some live in the water. Many salamanders are active at night.

fire salamander

Carpathian newt

Like other amphibians, salamanders do some of their breathing through their skin. More than 400 species of salamanders have no lungs!

A few types of salamanders do not lay eggs. Instead, they give birth to larvae.

Northwestern salamanders attach their eggs to blades of pond grass.

Bigger Than You

The Chinese giant salamander is the world's largest amphibian. It can grow to nearly six feet (2 meters) long.

Newts are a type of salamander. Adult newts live in the water. Most have webbed feet. A paddle-like tail helps them swim.

Wild alpine newts can live for up to 20 years.

At Risk

Close to one-third of the world's amphibians are at risk of dying out. The main causes include climate change, pollution, and disease.

Salamanders come out of their hiding places to hunt for prey.

This is different from other salamanders. Other adult salamanders live mostly on land. They have toes for digging. Their tails are longer and more rounded. Young newts are called efts. They spend time on land.

This juvenile eastern newt will return to the water when it becomes an adult.

Caecilians

Caecilians (si-SIL-yuhns) are the third order of amphibians. They have no legs. Their bodies are long and snake-like. Most live underground. They have hard heads to help them **burrow**. They can't see well and can't hear. They have two **tentacles** to help them sense their surroundings. Most caecilians have needle-like teeth.

There are about 160 types of caecilians. They live in the tropics in Africa, South Asia, and South America. They rarely come to the surface. This makes them hard to study.

caecilian

ACTIVITY

Candy Classification

Use candy to try your hand at creating your own classification system!

Supplies

index cards

string

tape

a bag with six to ten kinds of candy

Directions

1. Look at the different types of candy in the bag. Choose one characteristic to divide the candy into two smaller groups. For example, is the candy hard or not?

2. Using the index cards, label one group "hard" and the other group "not hard."

3. Now look at the hard candy group. What's another characteristic you can use to divide this group into two smaller groups? Label each group with an index card, and connect each of these new cards to the card before it with string and tape.

4. Keep dividing the candy into smaller and smaller groups, one characteristic at a time. Label each level, using the string to show where the smaller groups came from.

5. Keep going until you have piles with only one type of candy in each.

6. When you are done, give a piece of candy to a friend and see if they can classify it using your cards.

Glossary

burrow (BUR-oh): to dig or live in a tunnel or hole

carnivores (KAHR-nuh-vors): animals that eat meat

conditions (kuhn-DISH-uhns): the general state of things

larvae (LAHR-vee): newly hatched animals that are in a form very different from their parents

metamorphosis (met-uh-MOR-fuh-sis): a series of changes some animals go through as they develop into adults

tentacles (TEN-tuh-kuhls) the flexible limbs of some animals, used to feel things

toxin (TAHK-sin): poisonous

vertebrates (VUR-tuh-brates): animals that have a backbone

Index

Show What You Know

1. Name the three main groups of amphibians.

2. What do frogs and salamanders have in common?

3. How do frogs and salamanders differ?

4. Why do scientists classify living things?

5. Which kingdom do amphibians belong to?

Further Reading

Royston, Angela, *Amphibians (Animal Classifications)*, Heinemann, 2015.

Lee, Sally, *Amphibians: A 4D Book (Little Zoologist)*, Pebble, 2018.

Berne, Emma Carlson, *Amphibians (My First Animal Kingdom Encyclopedias)*, Capstone Press, 2017.

About the Author

Jodie Mangor writes magazine articles and books for children. She is also the author of audio tour scripts for high-profile museums and tourist destinations around the world. Many of these tours are for kids. She lives in Ithaca, New York, with her family.

Meet The Author!
www.meetREMauthors.com

www.rourkeeducationalmedia.com

PHOTO CREDITS: Cover & Title Page ©Wel_nofri; Border bgfoto; Pg 3 ©ivkuzmin; Pg 4 ©Mark Kostich; Pg 5 ©wiki, ©lvcandy; Pg 6 ©VvoeVale, ©Techin24: Pg 7 ©3drenderings; Pg 8 ©GlobalP; Pg 9 ©GlobalP; Pg 10 ©balipadma, Pg 11 ©marefoto, ©Wirepec; Pg 12 ©inhauscreative; Pg 13 ©MediaProduction; Pg 14 ©RolfAasa; Pg 15 ©Froggydarb; Pg 16 ©GlobalP, ©VitalisG; Pg 17 ©lisad1724; Pg 18 ©CreativeNature_nl, ©vectorplusb; Pg 19 ©Cristo Vlahos, ©popovaphoto; Pg 20 ©Hailshadow, Pg 22 ©GlobalP

Edited by: Keli Sipperley
Cover and interior design by: Kathy Walsh

Library of Congress PCN Data

Amphibians / Jodie Mangor
(Animals Have Classes Too!)
 ISBN 978-1-64369-030-8 (hard cover)
 ISBN 978-1-64369-108-4 (soft cover)
 ISBN 978-1-64369-177-0 (e-Book)
Library of Congress Control Number: 2018955964

Rourke Educational Media
Printed in the United States of America,
North Mankato, Minnesota